HANDBOOK OF STARSPOT GIANTS

The Magnetic Forces Behind Stellar Activity

MARCUS T. HOOKS

COPYRIGHT

Copyright©2024 Marcus T. Hooks. All rights reserved. No part of this publication may be reproduced, distributed, or transmitted in any form or by any means, including photocopying, recording, or other electronic or mechanical methods, without the prior written permission of the publisher, except in the case of brief quotations embodied in critical reviews and certain other non-commercial uses permitted by copyright law

TABLE OF CONTENTS

COPYRIGHT 1

TABLE OF CONTENTS 2

INTRODUCTION Handbook of Starspot Giants 4

CHAPTER 1 10

    The Discovery of XX Trianguli 10

CHAPTER 2 19

    Understanding Red Giant Stars 19

CHAPTER 3 30

    Chapter 3: What Are Starspots? 30

CHAPTER 4 43

    The Colossal Starspots of XX Trianguli 43

CHAPTER 5 54

    The Science Behind Starspot Formation 54

CHAPTER 6 67

    Mapping the Stellar Surface 67

CHAPTER 7 78

    Magnetic Dynamos and Stellar Activity 78

CHAPTER 8 92

The Impact of Stellar Activity on Surrounding Systems 92

CHAPTER 9 104

Implications for Stellar Evolution............................... 104

CHAPTER 10 117

Exploring the Universe Through Starspots.................. 117

CONCLUSION.. 129

Handbook of Starspot Giants 129

INTRODUCTION
Handbook of Starspot Giants

The universe is a tapestry of astonishing phenomena, each unraveling stories of creation, transformation, and cosmic evolution. Among the many mysteries that fill the celestial expanse, one particularly captivating enigma is the presence of starspots on red giant stars. These colossal blemishes, often larger than our own Sun, challenge our understanding of stellar dynamics, magnetic fields, and the life cycle of stars. *Handbook of Starspot Giants* delves into this fascinating topic, combining cutting-edge research, accessible explanations, and the awe-inspiring wonders of astrophysics.

A Glimpse into the Giants

Red giant stars are among the most majestic and enigmatic objects in the universe. These aging stars represent a phase of stellar evolution where ordinary stars, like our Sun, transform into luminous, bloated behemoths nearing the end of their lives. Yet, within their grandeur lies a curious feature: starspots—vast regions of magnetic activity where temperatures drop, creating cooler, darker patches on their fiery surfaces. These spots are not mere blemishes but gateways to understanding the intricate workings of magnetic fields, stellar rotation, and even the future of our own Sun.

Starspots on red giants differ significantly from those seen on smaller stars like our Sun. While solar spots are generally transient and relatively small, starspots on red giants are astonishingly massive, sometimes spanning billions of kilometers. They exhibit unique behaviors influenced by the star's immense size, slow rotation, and complex magnetic activity. What causes these spots? Why are they so massive? And what do they tell us about the life and death of stars? This handbook aims to explore these questions and more.

Why Starspots Matter

At first glance, starspots might seem like a minor curiosity, an imperfection in the otherwise dazzling glow of a star. However, their importance extends far beyond their appearance. Starspots offer clues about a star's magnetic field, rotation rate, and internal dynamics. They help scientists decipher the history of a star, its age, and its future evolution. For astronomers, they are a cosmic Rosetta Stone, providing insights into processes that are otherwise hidden beneath the seething plasma of a star's surface.

Moreover, studying starspots on red giants has profound implications for understanding stellar magnetism—a force that shapes not only stars but also the environments of their surrounding planets. Magnetic fields influence stellar winds, which, in turn, affect the habitability of planets in a star's system. By examining the interplay between starspots and stellar magnetism, we gain a deeper understanding of the forces that govern the universe and our place within it.

Bridging the Gaps in Knowledge

Despite their importance, starspots on red giants remain a relatively underexplored area of astrophysics. Advances in observational technology, including space telescopes and spectropolarimetry, have allowed scientists to detect and study these massive spots with unprecedented clarity. Yet, many questions linger. How do these spots form? What sustains them? And how do they influence the star's evolution and eventual fate?

This book seeks to bridge the gaps in our knowledge, offering a comprehensive overview of what we know—and what we don't—about starspot giants. Drawing from the latest research, it synthesizes information from various disciplines, including stellar astrophysics, magnetohydrodynamics, and observational astronomy.

Whether you are a professional astrophysicist, a student of the stars, or simply a curious stargazer, this handbook will guide you through the mesmerizing world of starspots.

The Journey Ahead

The structure of this book reflects the multifaceted nature of the topic. We begin with an overview of red giant stars, tracing their evolution from main-sequence stars to their current state as swollen, luminous giants. This sets the stage for a deeper exploration of starspots, their formation, and their characteristics. We'll dive into the physics of magnetism, unraveling the forces that shape these cosmic phenomena. The book also examines the observational techniques used to study starspots, from Doppler imaging to interferometry, highlighting the technological innovations that have revolutionized our understanding.

In later chapters, we'll explore the broader implications of starspots on stellar systems, including their impact on stellar winds, planetary habitability, and the star's ultimate demise. Finally, we'll look ahead to future research, identifying the key questions and challenges that will shape the field in the years to come.

Why This Handbook?

Astrophysics can often seem like an abstract science, filled with equations and theories that feel distant from everyday experience. Yet, at its heart, it is a profoundly human endeavor—a quest to understand the universe and our place within it. *Handbook of Starspot Giants* is designed to be both scientifically rigorous and accessible, balancing technical detail with clear explanations and vivid imagery. It invites readers to marvel at the cosmos, to ask questions, and to appreciate the beauty and complexity of the universe.

This handbook is also a tribute to the collaborative nature of science. The study of starspots on red giants is not the work of a single individual or discipline but a collective effort involving astronomers, physicists, engineers, and data scientists. It is a story of innovation and discovery, driven by the desire to push the boundaries of knowledge.

Looking to the Stars

As we peer into the heavens, we are reminded of the vastness of the cosmos and the smallness of our place within it. Yet, it is precisely this contrast that makes our exploration so meaningful. By studying phenomena like starspots, we not only unravel the secrets of distant stars but also gain insights into the forces that shaped our own solar system. We learn about the past, present, and future of the cosmos, and in doing so, we deepen our connection to the universe.

The red giants with their colossal starspots are not just objects of scientific inquiry—they are symbols of transformation and endurance. They teach us about cycles of growth and decline, about the interplay of light and darkness, and about the interconnectedness of all things. As you journey through this handbook, I invite you to reflect on the profound lessons these celestial giants offer.

A Note to the Reader

Whether you are a seasoned astronomer or a curious enthusiast, this book is your companion in exploring the mysterious world of starspot giants. It is a guide, a resource, and a celebration of the wonders of the universe. May it inspire you to look to the stars with a sense of wonder and curiosity, and may it deepen your appreciation for the cosmic dance of creation and transformation.

Let us now embark on this journey into the heart of the giants—into the mesmerizing world of starspots, where science meets wonder, and the cosmos reveals its secrets.

CHAPTER 1

The Discovery of XX Trianguli

"To confine our attention to terrestrial matters would be to limit the human spirit." — Stephen Hawking

The story of XX Trianguli begins with an observation that would shift our understanding of the dynamic lives of stars. Nestled within the constellation of Triangulum, this seemingly unremarkable red giant star became a focal point for astronomers, sparking intrigue with its peculiar behavior. From the moment it was first studied, XX Trianguli has served as a cosmic laboratory for understanding stellar evolution, magnetic activity, and the phenomena of starspots—gargantuan blemishes on a star's surface that dwarf anything seen in our solar system.

This chapter chronicles the discovery of XX Trianguli, the reasons it captured the attention of the astronomical community, and its role as an exemplary subject for studying red giant stars. To fully appreciate its significance, we first delve into the basics of red giant stars, their unique features, and why they are pivotal in understanding the life cycles of stars.

The Initial Observations and Why the Star Caught Astronomers' Attention

Early Surveys and XX Trianguli

In the late 20th century, as astronomers refined their tools for observing distant stars, XX Trianguli emerged as an intriguing outlier. It was first cataloged in a routine stellar survey, part of efforts to map stars in the Triangulum constellation. At the time, astronomers were employing a combination of optical and spectroscopic techniques to classify stars based on their luminosity, temperature, and composition.

XX Trianguli's peculiarities became evident during follow-up observations aimed at measuring its variability. It exhibited pronounced fluctuations in brightness, an indicator that the star was not stable but undergoing dynamic processes. The initial assumption was that it might be a pulsating variable star—a common phenomenon among red giants. However, further data revealed something far more complex.

When astronomers examined the star using spectroscopic analysis, they noticed anomalies in its spectral lines. These anomalies suggested the presence of large, cooler regions on its surface, akin to the sunspots on our own Sun but on an unimaginably larger scale. These starspots, later confirmed through advanced techniques such as Doppler imaging, were unlike anything previously observed in a star of this type.

The Role of Modern Telescopes

The discovery of XX Trianguli's starspots owes much to the advancements in observational astronomy. Telescopes equipped with spectropolarimeters allowed researchers to study the magnetic fields of stars, revealing that XX Trianguli possessed an unusually strong and complex magnetic field for a red giant. This was unexpected, as red giants are typically characterized by slow rotation and weak magnetic activity.

Another breakthrough came with the use of space-based telescopes, such as the Kepler Space Telescope, which could monitor the star's brightness variations with incredible precision. The periodic dimming patterns observed in XX Trianguli provided additional evidence of starspots and offered clues about the star's rotation period.

What made XX Trianguli even more compelling was its proximity—relatively speaking—in astronomical terms. Located approximately 300 light-years away, it was close enough for detailed study but far enough to avoid interference from interstellar dust. This balance of accessibility and mystery made it a prime candidate for ongoing research.

Introduction to Red Giant Stars and Their Unique Features

The Life Cycle of a Star

To understand the significance of XX Trianguli, it's essential to place it within the broader context of stellar evolution. Stars, including our Sun, begin their lives as dense clouds of gas and dust. Over millions of years, gravitational forces compress this material into a core where nuclear fusion ignites. This fusion process converts hydrogen into helium, releasing energy that causes the star to shine.

For most of its life, a star remains in the main sequence stage, burning hydrogen in its core. However, as the hydrogen supply diminishes, the star undergoes dramatic changes. It expands and cools, becoming a red giant. This phase is characterized by a bloated, luminous outer layer and a dense core where helium fusion eventually begins.

Red giant stars are among the most striking celestial objects, both in terms of their appearance and their significance. They represent the aging population of stars, offering a glimpse into the future of our Sun and other main-sequence stars. But they are also turbulent, dynamic entities, marked by complex internal processes and dramatic surface phenomena.

Characteristics of Red Giant Stars

1. **Enormous Size and Luminosity**

   Red giants are aptly named for their colossal size. A typical red giant can expand to hundreds of times the radius of the Sun, with a corresponding increase in brightness. This expansion occurs because the star's outer layers are no longer held tightly by gravity, allowing them to balloon outward. XX Trianguli, for instance, has a radius approximately 150 times that of the Sun and shines thousands of times brighter.

2. **Cooler Surface Temperature**

   Despite their brilliance, red giants have relatively low surface temperatures compared to their main-sequence counterparts. Their outer layers cool as they expand, giving them their characteristic reddish hue. For XX Trianguli, this cooler temperature made the starspots particularly noticeable, as the contrast between the spots and the surrounding surface was more pronounced.

3. **Unstable Interiors and Magnetic Fields**

   Red giants are inherently unstable. The lack of a stable core and the interplay of convection currents in their outer layers create turbulent conditions. This turbulence, combined with the star's magnetic activity, leads to the formation of starspots.

Magnetic fields in red giants are particularly intriguing. While magnetic activity typically diminishes as stars age and slow their rotation, some red giants, like XX Trianguli, defy this trend. The mechanisms driving their magnetic fields are not yet fully understood, making them a key area of study.

4. **Variable Brightness**

   Many red giants exhibit variability in their brightness, either due to pulsations or surface phenomena like starspots. XX Trianguli's brightness fluctuations were one of the first indicators of its unique characteristics.

Why Red Giants Matter

Red giant stars are more than just an intermediate phase in the life cycle of a star. They are cosmic alchemists, responsible for creating and dispersing heavy elements like carbon, nitrogen, and oxygen into the universe. These elements, forged in the hearts of stars, are the building blocks of planets, life, and everything we see around us.

Studying red giants helps astronomers understand the processes of stellar nucleosynthesis and the recycling of matter in the cosmos. They also provide a glimpse into the future of our own Sun, which will eventually become a red giant. By studying stars like XX Trianguli, we gain insights into the forces that will shape the solar system billions of years from now.

XX Trianguli: A Case Study in Stellar Complexity

The discovery of starspots on XX Trianguli challenged many assumptions about red giants. Traditionally, starspots were associated with younger, more active stars, where rapid rotation and strong magnetic fields drive the formation of spots. The presence of such massive starspots on XX Trianguli raised new questions about the nature of magnetic activity in red giants.

Starspot Giants

Starspots on XX Trianguli are among the largest ever recorded. Some spots cover areas hundreds of times the surface area of Earth, with temperatures thousands of degrees cooler than the surrounding plasma. These spots form when magnetic field lines on the star's surface become twisted and inhibit the flow of heat.

One of the most remarkable aspects of XX Trianguli's starspots is their longevity. Unlike the transient sunspots on our Sun, which typically last days or weeks, the spots on XX Trianguli can persist for months or even years. This persistence suggests that the magnetic fields driving their formation are highly stable, an unusual trait for a red giant.

Implications for Stellar Magnetism

The study of XX Trianguli has provided new insights into the magnetic fields of aging stars. It has shown that even slow-rotating stars can sustain complex magnetic activity, challenging previous models of stellar magnetism. Researchers have proposed several theories to explain this, including the possibility of a hidden dynamo effect within the star's core.

The Broader Context

XX Trianguli is more than just an astronomical curiosity—it is a gateway to understanding the complex interplay of forces that shape stars and their evolution. Its discovery has inspired new lines of inquiry, from the dynamics of magnetic fields to the mechanisms of starspot formation.

But perhaps most importantly, XX Trianguli reminds us of the beauty and complexity of the universe. It is a testament to the fact that even in the vastness of space, where billions of stars shine, each one has a story to tell.

As we journey deeper into the mysteries of XX Trianguli and its starspots, we move closer to answering some of the most profound questions in astrophysics. Why do stars behave the way they do? What forces govern their lives and deaths? And what can they teach us about our own place in the cosmos?

CHAPTER 2

Understanding Red Giant Stars

"Stars are the beacons of the universe, lighting the path to understanding the cosmos and our place within it."

In the grand tapestry of the cosmos, stars play a central role, not only as sources of light and energy but as engines of creation and transformation. Among the myriad types of stars that populate the universe, red giants stand out as particularly fascinating and significant. They are luminous, vast, and often tempestuous, representing a late and dramatic stage in the life cycle of stars. For XX Trianguli, a remarkable red giant nestled within the constellation of Triangulum, these characteristics are magnified by its extraordinary features.

This chapter explores the life cycle of stars, the processes that lead to the formation of red giants, and how XX Trianguli compares to its celestial counterparts. By understanding the unique traits of red giant stars, we can better appreciate their role in the evolution of the universe and their contribution to the ongoing mystery of stellar phenomena.

The Life Cycle of Stars and the Formation of Red Giants

The Birth of a Star

Stars begin their lives as clouds of gas and dust, known as nebulae. These nebulae, often located in star-forming regions of galaxies, are composed primarily of hydrogen, with traces of helium and heavier elements. Under the influence of gravity, these clouds begin to contract, becoming denser and hotter over millions of years. As the core temperature reaches approximately 10 million degrees Kelvin, nuclear fusion ignites.

In the fusion process, hydrogen nuclei (protons) combine to form helium, releasing vast amounts of energy in the form of light and heat. This energy counterbalances the inward pull of gravity, creating a stable structure—a main-sequence star.

The duration of the main-sequence phase depends on the star's mass. Smaller stars, like red dwarfs, can remain in this stage for tens of billions of years, while massive stars burn through their fuel in just a few million years. Our Sun, a mid-sized star, has been in the main-sequence phase for approximately 4.6 billion years and is expected to remain so for another 5 billion years.

The Transition to Red Giants

As a star exhausts the hydrogen in its core, the balance between gravity and nuclear fusion begins to break down. The core contracts under gravity, becoming hotter and denser. Meanwhile, the outer layers expand dramatically, cooling as they spread out. This marks the transition to the red giant phase.

Key processes during this transformation include:

1. **Helium Fusion (Helium Flash)**

   Once the core temperature rises to around 100 million degrees Kelvin, helium nuclei begin to fuse into carbon and oxygen. This process, known as the helium flash, releases an immense amount of energy, temporarily stabilizing the star.

2. **Convective Envelopes**

   The outer layers of a red giant become convective, meaning energy is transported by the movement of plasma rather than radiation. This convective activity plays a crucial role in shaping the star's surface features, including starspots.

3. **Mass Loss**

   Red giants lose significant amounts of mass through stellar winds. This material, rich in heavier elements, is returned to the interstellar medium, contributing to the formation of new stars and planets.

For stars with masses similar to the Sun, the red giant phase is a prelude to their final stages as white dwarfs. However, for more massive stars, this phase can lead to supernovae and the formation of neutron stars or black holes.

The Defining Characteristics of Red Giants

1. **Size and Luminosity**

   Red giants are among the largest stars in the universe, with radii ranging from tens to hundreds of times that of the Sun. Their luminosity is similarly vast, often thousands of times greater than the Sun, despite their cooler surface temperatures.

2. **Surface Temperature and Color**

   The cooler outer layers of red giants give them their characteristic reddish hue. Surface temperatures typically range from 2,500 to 5,000 Kelvin, much cooler than the Sun's 5,778 Kelvin.

3. **Unstable Interiors**

   The core of a red giant is a dense, contracting sphere surrounded by layers of hydrogen and helium undergoing fusion. This layered structure creates a dynamic, unstable environment, leading to pulsations and surface activity such as starspots.

4. **Chemical Enrichment**

   Red giants play a key role in enriching the universe with heavier elements. Fusion processes within the star produce carbon, oxygen, and other elements, which are released into space when the star sheds its outer layers.

The Unique Journey of XX Trianguli

XX Trianguli fits the profile of a red giant but stands out due to several unusual characteristics. Its extraordinary starspots, magnetic activity, and variability make it an exceptional case study. Understanding XX Trianguli requires comparing it to other red giants, highlighting both commonalities and distinctions.

How XX Trianguli Compares to Other Red Giants

A Typical Red Giant vs. XX Trianguli

1. **Size and Mass**

    XX Trianguli, like other red giants, has expanded to an enormous size, with a radius approximately 150 times that of the Sun. However, its mass is slightly lower than other red giants of comparable size, a fact that may contribute to its unusual magnetic activity.

2. **Magnetic Activity and Starspots**

    While most red giants exhibit weak magnetic fields due to their slow rotation, XX Trianguli is an anomaly. Its strong and complex magnetic field supports the formation of colossal starspots, some of which cover up to 20% of the star's surface. These spots are stable and persistent, lasting far longer than typical solar spots.

3. **Variability**

    Many red giants are variable stars, meaning their brightness fluctuates over time. This variability is often caused by pulsations in the star's outer layers or by the presence of starspots. XX Trianguli's brightness variations are particularly pronounced, with periodic dimming that aligns with its rotation period.

Starspot Giants: A Rare Phenomenon

The starspots on XX Trianguli are among the largest ever observed, dwarfing those on other red giants. For comparison:

- **Aldebaran**, a well-known red giant in the Taurus constellation, exhibits minor surface activity but lacks the massive starspots seen on XX Trianguli.

- **Betelgeuse**, another iconic red giant, has shown surface irregularities that may be linked to magnetic activity, but its starspots are transient and less well-defined.

XX Trianguli's starspots are remarkable not only for their size but also for their persistence and the insights they provide into the star's internal dynamics.

Rotation and Magnetic Fields

The rotation of a star plays a crucial role in generating its magnetic field. In main-sequence stars, the combination of rapid rotation and convection drives a dynamo effect, creating strong magnetic fields. However, red giants typically rotate more slowly, and their dynamo activity diminishes.

XX Trianguli is a notable exception. Despite its slow rotation, it exhibits a magnetic field comparable to that of younger, more active stars. Researchers speculate that this may be due to internal processes, such as the interaction between the core and the convective envelope, or residual angular momentum from its earlier stages.

Chemical Composition and Enrichment

Like other red giants, XX Trianguli is enriched with heavier elements produced during fusion. Spectroscopic studies reveal elevated levels of carbon, oxygen, and nitrogen in its outer layers, indicating active nucleosynthesis.

However, XX Trianguli also shows unusual spectral signatures that hint at unique chemical processes. For example, it has an overabundance of certain isotopes, suggesting that its fusion processes or mass-loss mechanisms may differ from those of typical red giants.

The Broader Implications of Studying Red Giants

Stellar Evolution and the Fate of the Sun

Studying red giants like XX Trianguli provides critical insights into the future of our Sun. Approximately 5 billion years from now, the Sun will exhaust its hydrogen fuel and begin its transformation into a red giant. It will expand to engulf the inner planets, including Earth, and shed its outer layers, leaving behind a white dwarf.

By observing stars like XX Trianguli, astronomers can model these processes with greater accuracy, predicting how the Sun will evolve and how its changes will impact the solar system.

The Role of Red Giants in Galactic Evolution

Red giants are key players in the cosmic cycle of matter. Their fusion processes produce the elements that form the building blocks of planets and life. When they shed their outer layers, they enrich the interstellar medium with these elements, fueling the next generation of star and planet formation.

XX Trianguli, with its pronounced mass loss and unique composition, offers a window into these processes. By studying its chemical output, astronomers can better understand how red giants contribute to the chemical evolution of galaxies.

Magnetic Fields and Stellar Dynamics

The magnetic activity of XX Trianguli challenges existing models of stellar magnetism. Understanding how such a strong magnetic field is sustained in a slowly rotating red giant could lead to breakthroughs in our knowledge of stellar dynamos and magnetohydrodynamics.

Exoplanetary Systems and Habitability

The behavior of red giants has significant implications for the planets that orbit them. As stars expand and emit stronger stellar winds, they can strip away planetary atmospheres or destabilize orbits. By studying XX Trianguli, researchers can assess how red giants affect their planetary systems, shedding light on the potential habitability of planets around aging stars.

The journey of a star from its birth in a nebula to its transformation into a red giant is a story of creation, change, and eventual renewal. Red giants like XX Trianguli are both the products of this journey and its most spectacular phase, offering a glimpse into the forces that shape the universe.

XX Trianguli stands out as a particularly fascinating example, defying expectations with its immense starspots, strong magnetic field, and unique chemical signatures. By comparing it to other red giants, we deepen our understanding of the diversity and complexity of stellar evolution.

CHAPTER 3

Chapter 3: What Are Starspots?

"The cosmos is within us. We are made of star-stuff. We are a way for the universe to know itself." — Carl Sagan

The surfaces of stars are dynamic, ever-changing regions where intense magnetic activity manifests in a variety of fascinating phenomena. Among these, starspots are one of the most intriguing. These colossal, cooler regions on a star's surface represent a fascinating intersection of astrophysics, magnetism, and thermodynamics. While sunspots on our own Sun have been studied for centuries, the discovery and examination of starspots on distant stars have significantly expanded our understanding of stellar behavior.

In this chapter, we will explore the nature of starspots, their formation, and their similarities to sunspots. We will also delve into the methods astronomers use to detect and study starspots on distant stars, providing insights into the tools and techniques that allow us to unravel these cosmic mysteries.

Explaining Starspots and Their Similarity to Sunspots

What Are Starspots?

Starspots are cooler, darker regions that form on the surface of stars due to intense magnetic activity. These regions appear darker because their temperatures are significantly lower than the surrounding areas of the star's surface. While the average temperature of a star's surface (photosphere) might be thousands of degrees Kelvin, starspots can be hundreds or even thousands of degrees cooler, resulting in their characteristic dim appearance when observed against the bright stellar background.

On our Sun, these regions are called sunspots. They appear as temporary blemishes on the Sun's photosphere and are associated with heightened magnetic activity. On other stars, similar phenomena are termed starspots, though they can differ significantly in size, structure, and longevity, especially on larger stars like red giants.

The Magnetic Origin of Starspots

Starspots are closely tied to the magnetic fields of stars. In the Sun, the magnetic field is generated by a process known as the solar dynamo, driven by the interaction between the Sun's rotation and the convective movement of plasma in its outer layers. This magnetic activity leads to the emergence of concentrated magnetic field lines at the surface, disrupting the normal convective flow of heat. These disruptions result in cooler, darker regions—sunspots.

The same principles apply to other stars. In stars with strong magnetic fields, regions of suppressed convection form, leading to the appearance of starspots. The size and distribution of starspots depend on factors such as the star's rotation rate, magnetic field strength, and age.

Similarities to Sunspots

Sunspots provide an accessible analogy for understanding starspots on distant stars. Key similarities include:

1. **Magnetic Activity:**

    Both sunspots and starspots are caused by intense magnetic fields that inhibit the upward flow of hot plasma from the star's interior. This results in cooler regions on the surface.

2. **Temperature Contrast:**

    Like starspots, sunspots are cooler than their surroundings, with temperatures typically around

3,000–4,500 Kelvin compared to the Sun's average surface temperature of about 5,778 Kelvin.

3. **Lifecycle:**

   Sunspots have lifespans ranging from days to months. Similarly, some starspots appear transient, while others, particularly on giant stars, can persist for years or even decades.

4. **Cycles:**

   Both sunspots and starspots often follow cyclical patterns. For the Sun, this is the 11-year solar cycle, during which the number and intensity of sunspots wax and wane. Many other stars exhibit similar periodic variations in magnetic activity.

How Are Starspots Different from Sunspots?

While the basic principles are the same, starspots can differ dramatically from sunspots in several ways:

1. **Size:**

   Starspots on certain stars, particularly red giants like XX Trianguli, can be colossal, spanning a significant fraction of the star's surface. Some starspots are thousands of times larger than the largest sunspots ever observed.

2. **Lifespan:**

   While sunspots typically last weeks or months, starspots on some stars can persist for years or even decades. This suggests differences in the underlying magnetic processes.

3. **Temperature Contrast:**

   The temperature difference between a starspot and its surrounding photosphere is often greater than that of sunspots. This is particularly pronounced on stars with cooler overall temperatures, such as red giants.

4. **Distribution:**

   Sunspots tend to cluster near the Sun's equator. On other stars, starspots can appear at higher latitudes, even near the poles, due to differences in magnetic field geometry.

5. **Multiplicity:**

   While the Sun typically has a limited number of sunspots at any given time, stars with strong magnetic activity may have multiple starspots distributed across their surfaces, often forming complex patterns.

Starspots on Different Types of Stars

Starspots are not uniform across all stars. Their characteristics vary depending on the type of star:

1. **Red Dwarfs:**

   These small, cool stars often exhibit intense magnetic activity due to their rapid rotation rates. Starspots on red dwarfs can cover a substantial fraction of the star's surface, making them appear dimmer overall.

2. **Young Stars:**

   Younger stars tend to rotate more rapidly, generating strong magnetic fields. As a result, they often have larger and more numerous starspots compared to older stars.

3. **Red Giants:**

   Starspots on red giants, such as XX Trianguli, are particularly intriguing. These spots can cover vast areas and persist for years, providing valuable insights into the magnetic activity of aging stars.

4. **Active Stars:**

   Some stars exhibit heightened magnetic activity, leading to the presence of numerous starspots and intense stellar flares. These stars are often referred to as magnetically active stars.

How Starspots Are Detected and Studied in Distant Stars

Detecting and studying starspots on distant stars presents unique challenges. Unlike sunspots, which can be directly observed using telescopes, starspots require indirect methods of detection due to the vast distances involved. Over the past several decades, astronomers have developed a range of sophisticated techniques to study these fascinating features.

Methods for Detecting Starspots

1. **Photometric Variations:**

   One of the primary ways astronomers detect starspots is by observing variations in a star's brightness over time. As a star rotates, starspots on its surface move in and out of view, causing periodic dimming and brightening. This method has been particularly effective with space-based telescopes like **Kepler** and **TESS**, which can monitor stars with incredible precision.

- **Light Curves:**

    The periodic dips and rises in a star's brightness are plotted as light curves. These curves provide information about the size, location, and rotation period of the starspots.

2. **Doppler Imaging:**

Doppler imaging is a powerful technique for creating detailed maps of a star's surface. It relies on the Doppler effect, where the motion of the star's surface creates shifts in the wavelength of light. By analyzing these shifts, astronomers can infer the presence of cooler regions (starspots) and map their distribution.

3. **Spectropolarimetry:**

Spectropolarimetry involves measuring the polarization of light emitted by a star. This technique is particularly useful for studying magnetic fields and can reveal the presence of starspots by detecting regions of strong magnetic activity.

4. **Interferometry:**

   Using interferometers, astronomers can combine light from multiple telescopes to achieve high-resolution images of nearby stars. This technique has been used to directly observe surface features, including starspots, on some stars.

5. **Asteroseismology:**

   By studying the oscillations or "starquakes" of a star, astronomers can gain insights into its internal structure and surface activity. These oscillations are affected by the presence of starspots, providing indirect evidence of their size and distribution.

What We Learn from Studying Starspots

1. **Stellar Rotation:**

   The movement of starspots across a star's surface provides a direct measure of the star's rotation period. This is particularly valuable for studying stars whose rotation cannot be measured through other means.

2. **Magnetic Fields:**

   Starspots offer clues about the strength and geometry of a star's magnetic field. By mapping starspots, astronomers can infer the structure of the star's magnetic dynamo.

3. **Stellar Cycles:**

Observing changes in starspots over time helps astronomers identify cycles of magnetic activity, similar to the Sun's 11-year solar cycle. This information is critical for understanding the long-term behavior of stars.

4. **Stellar Age:**

The presence and characteristics of starspots can provide indirect clues about a star's age. Younger stars tend to have more starspots due to their rapid rotation, while older stars show fewer spots as their magnetic activity diminishes.

5. **Planetary Impacts:**

Starspots can have significant effects on the environments of planets orbiting a star. For example, intense magnetic activity associated with starspots can produce stellar flares and winds that influence planetary atmospheres.

Technological Advances in Starspot Research

The study of starspots has been revolutionized by advancements in technology. Space-based observatories, such as the **Kepler Space Telescope**, have enabled astronomers to monitor thousands of stars simultaneously, providing an unprecedented dataset for studying starspots. Upcoming missions, like the **PLATO (PLAnetary Transits and Oscillations of stars)** mission, promise to further enhance our understanding.

Ground-based facilities, such as the **European Southern Observatory (ESO)** and the **Atacama Large Millimeter/submillimeter Array (ALMA)**, also play a vital role. High-resolution spectrographs and interferometers allow astronomers to observe starspots with increasing precision.

Case Study: XX Trianguli and Its Starspots

XX Trianguli provides an exceptional case study for the study of starspots. Its enormous size as a red giant, combined with its strong magnetic field, has led to the formation of some of the largest and most persistent starspots ever observed. By applying the techniques outlined above, astronomers have been able to map the distribution of these starspots and gain insights into the unique magnetic processes at work in red giants.

The Broader Implications of Studying Starspots

Starspots are not merely cosmetic features of stars—they hold the keys to understanding a wide range of astrophysical phenomena. From the magnetic fields that shape stellar behavior to the effects of stellar activity on planetary systems, the study of starspots offers insights into the fundamental workings of the universe.

Moreover, by studying starspots across a diverse range of stars, astronomers can develop models that explain the evolution of magnetic activity over a star's lifetime. These models have implications not only for astrophysics but also for understanding the long-term habitability of planets, including those in our own solar system.

Starspots are one of the most fascinating manifestations of stellar magnetism, offering a window into the dynamic processes that govern stars. From their formation and characteristics to the methods used to study them, starspots reveal a wealth of information about the life and behavior of stars. For stars like XX Trianguli, starspots are not just features—they are a defining characteristic that challenges our understanding of stellar evolution and magnetism.

As technology advances and our observational capabilities grow, the study of starspots will continue to illuminate the mysteries of the cosmos, bridging the gap between the stars and our understanding of their inner workings.

CHAPTER 4
The Colossal Starspots of XX Trianguli

"The stars seem eternal, but even in their steady glow lies the unpredictable rhythm of magnetic forces and the mysteries of cosmic scales."

XX Trianguli, a striking red giant in the constellation Triangulum, has drawn the attention of astronomers for reasons beyond its luminous brilliance. Its defining feature lies in the colossal starspots that mar its surface, offering a unique glimpse into the magnetic phenomena that occur on aging stars. These starspots are unlike anything observed on our Sun, both in scale and behavior. Their sheer size and persistence have not only defied prior expectations of red giant behavior but have also opened new doors in the study of stellar magnetism and evolution.

In this chapter, we will explore the colossal starspots of XX Trianguli, comparing their sizes to features on the Sun, and examining the unique properties that set them apart. By understanding what makes these starspots exceptional, we uncover the mechanisms driving them and their implications for the broader understanding of stellar activity.

Description of the Massive Starspots and Their Sizes Relative to the Sun

The Scale of XX Trianguli's Starspots

Starspots on XX Trianguli are astronomical phenomena of unprecedented scale. Observations reveal that these spots often cover vast portions of the star's surface, dwarfing sunspots by many orders of magnitude. While the largest sunspots observed on the Sun rarely exceed 50,000 kilometers in diameter, the starspots on XX Trianguli span hundreds of millions of kilometers—some nearly rivaling the size of the Sun itself.

For context:

- **The Sun's Largest Sunspots:**

  The Sun's largest sunspots, which occur during periods of peak solar activity, typically cover less than 0.3% of the Sun's surface area.

- **XX Trianguli's Starspots:**

  Starspots on XX Trianguli can cover up to **20–30% of the star's visible surface**. Given XX Trianguli's immense radius (approximately 150 times that of the Sun), the actual physical area of these starspots is staggering. A single starspot could encompass more than 10 billion square kilometers—an area large enough to fit multiple Earths.

A Comparison in Temperature and Brightness

Starspots, like sunspots, are cooler regions on the star's surface. On XX Trianguli, the average surface temperature is around 3,500 Kelvin. In contrast, the starspots can have temperatures as low as 2,500 Kelvin, creating a pronounced contrast between the dark spots and the surrounding luminous surface.

By comparison:

- **Sunspots:**

    Sunspots on our Sun exhibit a temperature drop of about 1,500 Kelvin, making them relatively cooler but still radiantly hot at approximately 4,000 Kelvin.

- **Starspots on XX Trianguli:**

    The temperature difference on XX Trianguli is even more significant, often exceeding 1,000 Kelvin. This sharp contrast makes the starspots distinctly visible through photometric observations and spectroscopic analysis, despite the overall brightness of the star.

The Persistence of Colossal Starspots

Another remarkable feature of XX Trianguli's starspots is their longevity. Unlike the transient sunspots on our Sun, which typically last for days or weeks, starspots on XX Trianguli can persist for **months or even years**. Their stability suggests underlying magnetic processes that are both strong and sustained—phenomena not commonly associated with red giants.

This persistence enables detailed studies of the star's rotation and magnetic field. As the star rotates, the movement of these massive starspots across its surface creates observable brightness variations, offering a window into the internal dynamics of the star.

What Makes XX Trianguli's Starspots Unique?

1. Extraordinary Size

The most striking aspect of XX Trianguli's starspots is their enormous size. These spots are orders of magnitude larger than any observed on smaller stars, including our Sun. The factors contributing to this extraordinary size include:

- **Slower Rotation:**

    Red giants like XX Trianguli rotate much more slowly than main-sequence stars. This slower rotation allows starspots to form and grow to immense sizes without the rapid shearing forces that might disrupt them on faster-rotating stars.

- **Larger Surface Area:**

    The vast surface area of XX Trianguli provides ample room for starspot formation. This is in contrast to smaller stars, where the available surface area limits the potential size of starspots.

- **Magnetic Field Strength:**

    Despite being a red giant, XX Trianguli exhibits an unusually strong magnetic field. This magnetic activity supports the formation of large, stable starspots.

2. High-Latitude Distribution

Unlike sunspots, which predominantly form near the Sun's equator, the starspots on XX Trianguli often appear at higher latitudes, including near the star's poles. This polar distribution is unusual and may be linked to the geometry of the star's magnetic field.

The magnetic field of XX Trianguli likely differs from the Sun's dipolar structure, which confines sunspots to lower latitudes. Instead, the field on XX Trianguli may be more complex, with multipolar components that allow starspots to form across a broader range of latitudes.

3. Longevity and Stability

The persistence of XX Trianguli's starspots is another unique characteristic. While sunspots are relatively short-lived due to the Sun's faster rotation and dynamic magnetic activity, the starspots on XX Trianguli endure for extended periods.

Several factors contribute to this stability:

- **Convective Zones:**

    Red giants have deep convective envelopes that play a significant role in sustaining magnetic fields. The interaction between these convective currents and the star's magnetic field may stabilize the starspots.

- **Magnetic Anchoring:**

    The strong magnetic fields on XX Trianguli may "anchor" starspots in place, preventing them from dissipating quickly.

4. Contribution to Brightness Variability

The size and distribution of starspots on XX Trianguli have a significant impact on the star's overall brightness. As the star rotates, the starspots move in and out of view, creating periodic variations in brightness.

This variability is much more pronounced than the brightness changes caused by sunspots on the Sun, which are relatively minor. On XX Trianguli, the presence of large, dark starspots can reduce the star's brightness by up to **10%** during peak activity periods.

5. The Role of Stellar Evolution

The presence of such massive starspots on XX Trianguli raises questions about the magnetic activity of red giants. Traditionally, red giants are expected to have weaker magnetic fields due to their slower rotation and expanded size. However, XX Trianguli defies this expectation, suggesting that some red giants can sustain strong magnetic activity even in their advanced stages of evolution.

This unique behavior may be linked to:

- **Internal Dynamo Effects:**

  The interaction between the contracting core and the expansive convective envelope in red giants may generate complex magnetic fields.

- **Legacy Rotation:**

  XX Trianguli may have retained some angular momentum from its earlier life stages, enabling it to sustain magnetic activity.

Case Study: Mapping the Starspots of XX Trianguli

Astronomers have used advanced techniques to map the starspots on XX Trianguli, providing unprecedented insights into their size, distribution, and behavior.

Doppler Imaging

Doppler imaging is a key technique for studying the surface features of stars. By analyzing the Doppler shifts in the star's light spectrum, astronomers can infer the presence of cooler regions on the surface. For XX Trianguli, Doppler imaging has revealed:

- **Massive Polar Spots:**

    Large, stable spots near the star's poles, covering significant portions of the star's visible surface.

- **Complex Spot Patterns:**

    Multiple starspots of varying sizes, distributed across different latitudes, suggesting a highly dynamic magnetic field.

Photometric Observations

Space-based telescopes like **Kepler** and **TESS** have monitored the brightness variations of XX Trianguli over time. These observations have:

- Confirmed the periodic dimming caused by starspot rotation.

- Provided data on the star's rotation period (approximately 400 days).

- Highlighted the long-term stability of the starspots.

Spectropolarimetry

Spectropolarimetry has been used to study the magnetic fields associated with XX Trianguli's starspots. This technique measures the polarization of light caused by magnetic activity, revealing:

- Strong, complex magnetic fields anchoring the starspots.
- Evidence of multipolar magnetic structures, which may explain the high-latitude distribution of the starspots.

Implications for Stellar Physics

The colossal starspots of XX Trianguli provide a unique opportunity to study the magnetic activity of red giants. Key implications include:

1. **Advancing Models of Stellar Magnetism:**

   The strong magnetic fields of XX Trianguli challenge existing models, which predict weaker magnetic activity in red giants. Studying these fields may lead to a better understanding of the dynamo processes in aging stars.

2. **Insights into Stellar Evolution:**

   By studying XX Trianguli, astronomers can gain insights into the late stages of stellar evolution, particularly the transition from red giant to white dwarf.

3. **Impact on Exoplanets:**

   The magnetic activity of XX Trianguli, including stellar winds and flares associated with starspots, could significantly impact any orbiting planets. Understanding this activity is crucial for assessing the habitability of planets around red giants.

The colossal starspots of XX Trianguli are a testament to the complexity and diversity of stellar behavior. Their immense size, persistence, and unique properties challenge our understanding of red giants and offer new avenues for research into stellar magnetism and evolution.

As we continue to study XX Trianguli, its starspots serve as a reminder of the dynamic and unpredictable nature of the universe. They are not merely blemishes on the surface of a star but key to unlocking the secrets of stellar physics and the forces that shape the cosmos.

CHAPTER 5
The Science Behind Starspot Formation

"Stars are furnaces of transformation, but their surfaces are battlegrounds of magnetic forces—revealing the hidden complexities within."

Starspots, one of the most captivating features of stellar surfaces, are visible manifestations of the magnetic activity within a star. From the small, transient sunspots on our Sun to the colossal, persistent starspots on XX Trianguli, these regions provide a window into the complex interplay of magnetic fields, convection, and rotation that shapes a star's behavior. Understanding starspot formation not only sheds light on the internal dynamics of stars but also highlights the diversity of stellar activity across the universe.

This chapter explores the science behind starspot formation, focusing on the role of magnetic fields in their creation and maintenance. It also compares the chaotic activity of XX Trianguli with the Sun's more predictable 11-year solar cycle, offering insights into the similarities and differences between these two stellar examples.

The Role of Stellar Magnetic Fields in Creating Starspots

The Basics of Magnetic Fields in Stars

Magnetic fields in stars arise from the movement of charged particles within their interiors. This motion is driven by a combination of rotation and convection, creating a dynamo effect. The dynamo is responsible for generating the magnetic fields that extend throughout the star and influence its surface phenomena, including the formation of starspots.

1. **The Dynamo Effect:**
    - The dynamo effect is powered by the interaction between the star's rotation and its convective zones.
    - In main-sequence stars like the Sun, the outer convective zone mixes plasma, while differential rotation (faster at the equator than at the poles) twists and stretches magnetic field lines.
    - The result is a complex, ever-changing magnetic field that cycles over time.

2. **Magnetic Field Lines and Surface Activity:**

    o Magnetic field lines emerge from the star's interior and penetrate its surface, forming regions of intense magnetic activity.

    o These regions suppress convection, reducing the upward flow of heat and causing localized cooling. This cooling results in the formation of dark spots—what we observe as starspots.

Starspot Formation: A Step-by-Step Process

1. **Magnetic Flux Emergence:**

    o Starspots begin deep within the star, where buoyant magnetic flux tubes form in the convection zone. These flux tubes, composed of tightly coiled magnetic field lines, are pushed toward the surface by the star's convective currents.

2. **Penetration of the Surface:**

    o When a flux tube reaches the star's photosphere (the visible surface), it breaks through, creating a concentrated region of magnetic activity.

3. **Suppression of Convection:**
   - In these regions, the magnetic field inhibits the normal convective flow of plasma, reducing the transport of heat from the star's interior to its surface.

4. **Temperature Drop and Darkening:**
   - The suppression of heat transfer leads to a temperature drop in the affected area, making it appear darker than the surrounding, hotter regions.

5. **Evolution and Decay:**
   - Starspots evolve as the magnetic field shifts and dissipates. Some may disappear within days, while others—particularly on stars like XX Trianguli—can persist for months or years.

The Role of Rotation and Convection

1. **Rotation Rate:**
   - The speed at which a star rotates has a significant impact on the formation of starspots. Faster rotation enhances the

dynamo effect, increasing magnetic activity and leading to more numerous or larger starspots.

- For XX Trianguli, the relatively slow rotation (approximately 400 days) would typically suggest weaker magnetic activity. However, its deep convective zones and unique internal dynamics compensate for this, sustaining strong magnetic fields.

2. **Convective Envelope:**
    - Red giants like XX Trianguli have extensive convective envelopes, where hot plasma rises and cooler plasma sinks. This large-scale motion amplifies magnetic field generation and contributes to the formation of massive starspots.

Starspot Giants vs. Sunspots

While the processes of starspot and sunspot formation share the same underlying principles, key differences arise due to variations in stellar size, rotation, and internal structure:

1. **Size:**
   - Starspots on XX Trianguli can cover millions of square kilometers, dwarfing even the largest sunspots.

2. **Magnetic Field Strength:**
   - The magnetic fields associated with XX Trianguli's starspots are significantly stronger and more complex than those of the Sun, likely due to the star's convective envelope and internal magnetic dynamo.

3. **Temperature Contrast:**
   - The temperature difference between XX Trianguli's starspots and its surrounding surface is greater than that of the Sun, making the spots more prominent.

4. **Persistence:**
   - Sunspots are relatively short-lived, while starspots on XX Trianguli can last for years, reflecting the stability of the star's magnetic fields.

Comparing XX Trianguli's Chaotic Activity to the Sun's 11-Year Cycle

The Sun's 11-Year Solar Cycle

The Sun's magnetic activity follows a well-defined, periodic cycle lasting approximately 11 years. This cycle governs the appearance and behavior of sunspots, as well as other phenomena such as solar flares and coronal mass ejections.

1. **Solar Minimum and Maximum:**
   - During the solar minimum, the Sun's magnetic activity is at its lowest, and few or no sunspots are visible.
   - At the solar maximum, magnetic activity peaks, and sunspots proliferate across the Sun's surface.

2. **Magnetic Reversals:**
   - At the end of each 11-year cycle, the Sun's magnetic field reverses polarity. This magnetic reset sets the stage for the next cycle.

3. **Sunspot Behavior:**
   - Sunspots typically appear near the Sun's equator and migrate toward the poles over the course of the cycle. This predictable pattern is known as the "butterfly diagram."

4. **Impact of the Solar Cycle:**
   - The solar cycle influences the Sun's luminosity, solar wind intensity, and radiation output, which in turn affect space weather and conditions on Earth.

The Chaotic Activity of XX Trianguli

In contrast to the Sun's orderly 11-year cycle, the magnetic activity of XX Trianguli is highly chaotic and less predictable. This disparity arises from fundamental differences in the structure and behavior of the two stars:

1. **Lack of a Defined Cycle:**
   - Unlike the Sun, XX Trianguli does not exhibit a regular magnetic cycle. Its starspots appear and evolve in a more erratic manner, likely due to the star's unique internal dynamics.

2. **Complex Magnetic Field Geometry:**
   - The Sun's magnetic field is primarily dipolar (with a north and south pole), which contributes to its predictable behavior. In XX Trianguli, the magnetic field is multipolar, with numerous poles and regions of activity. This complexity leads to irregular starspot formation and distribution.

3. **Persistent Starspots:**
   - While sunspots are transient, starspots on XX Trianguli can persist for years. This longevity reflects the stability of the star's magnetic fields and the slower rate of flux dissipation.

4. **Impact of Stellar Evolution:**
   - As a red giant, XX Trianguli is in a later stage of stellar evolution than the Sun. Its core has contracted, and its outer layers have expanded, altering the behavior of its magnetic field and creating conditions conducive to chaotic activity.
   -

What XX Trianguli Teaches Us About Magnetic Activity

Despite their differences, XX Trianguli and the Sun share some commonalities that provide valuable insights into stellar magnetism:

1. **Universal Dynamo Principles:**
    - Both stars generate magnetic fields through dynamo processes, highlighting the universality of this mechanism across different types of stars.

2. **Magnetic Field Evolution:**
    - The chaotic activity of XX Trianguli offers a glimpse into how magnetic fields evolve as stars age and transition into later stages of their life cycle.

3. **Implications for Stellar Variability:**
    - Understanding the differences between XX Trianguli and the Sun underscores the diversity of stellar variability and the factors that influence it, such as mass, rotation, and composition.

Technological Advances in Studying Magnetic Fields and Starspots

The study of stellar magnetic fields and starspots has been revolutionized by advancements in observational techniques:

1. **Space-Based Telescopes:**
    - Missions like **Kepler** and **TESS** have provided high-precision photometric data, enabling the detection of brightness variations caused by starspots.

2. **Doppler Imaging:**
    - By analyzing the Doppler shifts in a star's spectrum, astronomers can create detailed maps of starspots and study their distribution and evolution.

3. **Spectropolarimetry:**
    - This technique measures the polarization of light caused by magnetic fields, offering insights into the strength and geometry of a star's magnetic field.
    -

4. **Asteroseismology:**
    - By studying stellar oscillations, astronomers can probe the internal structures of stars and gain a deeper understanding of the processes driving magnetic activity.

The Broader Implications of Studying Starspot Formation

The study of starspots extends beyond understanding individual stars like the Sun and XX Trianguli. It has broader implications for astrophysics, planetary science, and our understanding of the universe:

1. **Stellar Evolution:**
    - Starspot formation provides clues about the internal dynamics and life cycles of stars, shedding light on how they evolve over billions of years.

2. **Planetary Environments:**
    - The magnetic activity associated with starspots influences stellar winds and radiation, which can impact the atmospheres and habitability of planets orbiting the star.

3. **Stellar Magnetism:**
    - By comparing starspots across different types of stars, astronomers can develop more comprehensive models of stellar magnetism and its role in shaping stellar behavior.

The science behind starspot formation reveals the intricate interplay of magnetic fields, convection, and rotation that drives the dynamic behavior of stars. While the Sun's 11-year cycle provides a relatively orderly framework for understanding magnetic activity, the chaotic behavior of XX Trianguli highlights the diversity and complexity of stellar phenomena.

By studying these differences, we not only deepen our understanding of individual stars but also gain insights into the universal processes that govern the cosmos. As observational techniques continue to advance, the mysteries of starspots—and the stars that host them—will continue to inspire and challenge our understanding of the universe.

CHAPTER 6
Mapping the Stellar Surface

"To explore the stars is to venture into a world of light and shadow, where science meets imagination, and advanced techniques make the invisible visible."

Mapping the surface of distant stars, especially features as intricate as starspots, is one of the most challenging tasks in astrophysics. Stars are so far away that they appear as point sources of light, even through the most powerful telescopes. Yet, by leveraging cutting-edge techniques like Doppler imaging, spectropolarimetry, and asteroseismology, astronomers have developed methods to probe the surfaces of stars and unveil their dynamic behaviors.

This chapter explores the innovative methods used to map stellar surfaces, focusing on the techniques that have been instrumental in studying starspots on stars like XX Trianguli. We also examine the challenges inherent in observing distant stars and highlight the technological breakthroughs that are expanding our understanding of stellar surfaces.

Techniques Used to Map the Starspots

Mapping the surface of stars is an indirect process that relies on analyzing changes in light, color, and spectral characteristics to infer surface features. Below, we examine the primary methods used to detect and map starspots.

1. Doppler Imaging

Doppler imaging is the cornerstone technique for mapping stellar surfaces, particularly for identifying and studying starspots. The method exploits the Doppler effect, which shifts the wavelength of light based on the relative motion of a star's surface regions as it rotates.

How It Works:

- As a star rotates, its surface exhibits Doppler shifts:
    - Regions moving toward us appear slightly blue-shifted (light waves compressed).
    - Regions moving away from us appear red-shifted (light waves stretched).
- Starspots disrupt this regular pattern. Since starspots are cooler and darker, they emit less light. As a result, their presence creates irregularities in the star's spectrum during its rotation.

- By analyzing these spectral distortions over time, astronomers can reconstruct a map of the star's surface, identifying the location, size, and temperature of starspots.

Applications on XX Trianguli:

Doppler imaging has been pivotal in studying XX Trianguli. The massive starspots on its surface create pronounced disruptions in its spectral lines, allowing astronomers to produce detailed surface maps. These maps have revealed:

- **Large polar starspots**, some spanning millions of square kilometers.
- Irregular distributions of spots, highlighting the complexity of the star's magnetic field.

2. Photometric Light Curve Analysis

Photometry involves measuring a star's brightness over time. When a star rotates, starspots on its surface cause periodic dimming and brightening as they move in and out of view.

How It Works:

- Astronomers plot the variations in brightness to produce a **light curve**.

- The shape and periodicity of the light curve provide information about the star's rotation rate and the distribution of starspots.
- By combining light curves with models of stellar rotation, researchers can infer the size and location of starspots.

Limitations:

While photometric analysis is effective for identifying starspots, it lacks the spatial resolution of Doppler imaging. However, it is a complementary technique, particularly for stars too faint or distant for spectroscopic methods.

3. Spectropolarimetry

Spectropolarimetry measures the polarization of light emitted by a star. Polarization is influenced by magnetic fields, making this technique ideal for studying the magnetic activity associated with starspots.

How It Works:

- Magnetic fields distort the way light waves oscillate, creating polarized light.
- By analyzing this polarization across different wavelengths, astronomers can map the star's

magnetic field and identify regions of strong magnetic activity, such as starspots.

Applications on XX Trianguli:

Spectropolarimetry has confirmed that the massive starspots on XX Trianguli are associated with strong, complex magnetic fields. These measurements have also revealed that the star's magnetic field is multipolar, with multiple magnetic poles contributing to its chaotic surface activity.

4. Interferometry

Interferometry combines the light collected by multiple telescopes to achieve higher resolution than any single telescope could provide. This technique is particularly useful for resolving surface features on nearby stars.

How It Works:

- Light waves from different telescopes are combined to create an interference pattern.
- By analyzing this pattern, astronomers can reconstruct images of the star's surface.
- Interferometry has been used to directly image large starspots on some nearby stars, though it is limited by the star's distance and brightness.

Potential for XX Trianguli:

While interferometry has not yet resolved individual starspots on XX Trianguli, future advancements in interferometric techniques may make this possible.

5. Asteroseismology

Asteroseismology studies the oscillations, or "starquakes," of a star. These oscillations are influenced by the star's internal structure and surface activity, including the presence of starspots.

How It Works:

- As a star oscillates, its surface vibrates in specific patterns. These vibrations affect the star's brightness and spectral lines.

- By analyzing these oscillation patterns, astronomers can infer the star's internal structure and map surface irregularities, such as starspots.

Contributions to Starspot Studies:

Asteroseismology provides indirect evidence of starspots by revealing disruptions in the star's oscillation patterns. For XX Trianguli, asteroseismic data has complemented other techniques, confirming the presence of large, stable starspots.

6. Advances in Space-Based Observations

Space-based telescopes like **Kepler** and **TESS** have revolutionized the study of starspots. These telescopes provide high-precision photometric data, enabling astronomers to detect even subtle variations in brightness caused by starspots.

For XX Trianguli, these observations have provided:

- Accurate measurements of its rotation period.
- Long-term monitoring of its starspot activity, revealing changes over time.

Challenges in Studying Distant Stars

Mapping the surfaces of stars like XX Trianguli is an extraordinary feat, but it comes with significant challenges. Below, we explore the primary obstacles and how they are being addressed.

1. Distance and Resolution

The vast distances between Earth and stars make it nearly impossible to resolve their surfaces directly. Even the largest telescopes can only capture stars as point sources of light.

Addressing the Challenge:

- Techniques like Doppler imaging and interferometry have been developed to overcome this limitation.

- Future projects, such as the **Extremely Large Telescope (ELT)**, aim to provide unprecedented resolution for studying stellar surfaces.

2. Signal-to-Noise Ratio

Stellar observations are often hindered by noise, including interference from Earth's atmosphere and limitations in instrument sensitivity. For faint stars like XX Trianguli, obtaining a clear signal is particularly challenging.

Breakthroughs:

- Space-based telescopes eliminate atmospheric interference, providing cleaner data.

- Advances in spectroscopic instruments, such as high-resolution spectrographs, have improved sensitivity and reduced noise.

3. Magnetic Complexity

The chaotic magnetic fields of stars like XX Trianguli complicate efforts to map their surfaces. Unlike the Sun's relatively orderly dipolar field, XX Trianguli exhibits multipolar fields that create irregular starspot patterns.

Approaches to Understanding Complexity:

- Improved computational models simulate the interaction between a star's magnetic field and its surface activity.

- Long-term monitoring allows astronomers to track changes in magnetic activity and identify patterns.

4. Limitations of Current Techniques

Each mapping technique has its limitations:

- Doppler imaging requires high-quality spectral data, which can be difficult to obtain for distant or faint stars.

- Interferometry is limited to nearby stars and struggles with stars that have irregular brightness.

Overcoming Limitations:

- Combining multiple techniques provides a more comprehensive view of stellar surfaces. For example, Doppler imaging and spectropolarimetry together can map both starspots and magnetic fields.

Breakthroughs in Stellar Surface Mapping

Despite these challenges, significant breakthroughs have advanced the field of stellar surface mapping.

1. High-Resolution Spectrographs

Instruments like the **HARPS (High Accuracy Radial velocity Planet Searcher)** spectrograph have enabled precise measurements of stellar spectra, providing the data needed for Doppler imaging.

2. Machine Learning and Artificial Intelligence

Machine learning algorithms are being used to analyze large datasets from telescopes, improving the detection and mapping of starspots.

3. Future Missions

Upcoming space missions, such as the **PLATO (PLAnetary Transits and Oscillations of stars)** mission, promise to provide even more detailed data on stellar surfaces and oscillations.

The Future of Stellar Mapping

As technology continues to advance, the ability to map stellar surfaces will improve dramatically. Innovations in instrumentation, computational modeling, and data analysis will allow astronomers to study stars in greater detail than ever before. For stars like XX Trianguli, this means uncovering new insights into the behavior of red giants and their magnetic activity.

Mapping the stellar surface of stars like XX Trianguli is a triumph of modern astrophysics, achieved through a combination of innovative techniques and advanced technology. From Doppler imaging to spectropolarimetry, these methods have unveiled the dynamic, complex surfaces of stars that were once thought to be beyond our reach.

The challenges of studying distant stars are formidable, but breakthroughs in instrumentation and data analysis are pushing the boundaries of what is possible. As we continue to refine these techniques, the mysteries of stars like XX Trianguli will come into even sharper focus, deepening our understanding of the universe and the forces that shape it.

CHAPTER 7
Magnetic Dynamos and Stellar Activity

"Stars are not just sources of light; they are dynamic systems powered by invisible engines that dictate their magnetic activity and behavior. The magnetic dynamo is the hidden architect of stellar evolution."

Magnetic fields are one of the most enigmatic and crucial forces shaping the behavior of stars. Central to the creation and evolution of these fields is the magnetic dynamo—a complex, dynamic process operating within a star. By converting kinetic energy into magnetic energy, dynamos produce the magnetic fields that influence everything from surface starspots to large-scale stellar winds. Understanding magnetic dynamos is fundamental to unlocking the mysteries of stellar activity, including the unique behaviors observed in stars like XX Trianguli.

In this chapter, we explore the nature of magnetic dynamos, how they differ between various types of stars, and the implications of XX Trianguli's highly irregular and non-periodic magnetic activity. We will delve into how the study of this red giant contributes to our broader understanding of magnetic fields in stars and their role in the evolution of stellar systems.

What Are Magnetic Dynamos?

Definition and Principles

A magnetic dynamo is a mechanism within a star that generates and sustains its magnetic field. This process operates by converting the motion of electrically conducting fluids—plasma within the star—into magnetic energy. The dynamo effect occurs in regions of the star where convective and rotational forces interact, amplifying weak initial magnetic fields through feedback loops.

Key Components of a Stellar Dynamo:

1. **Convection:**

    The movement of hot plasma rising and cooler plasma sinking in a star's outer layers creates electric currents, which induce magnetic fields.

2. **Differential Rotation:**

    Stars typically rotate at different rates depending on latitude, with equatorial regions rotating faster than polar regions. This differential rotation twists magnetic field lines, amplifying the overall field.

3. **Feedback Loop:**

   The interaction between convection and rotation sustains the magnetic field, converting kinetic energy into magnetic energy.

4. **Magnetic Buoyancy:**

   Concentrated magnetic field lines rise to the surface, where they create starspots and other magnetic phenomena.

The Sun's Magnetic Dynamo: A Benchmark

The Sun provides the best-studied example of a magnetic dynamo. Its dynamo operates in the boundary region between the radiative core and the outer convective zone, known as the **tachocline**. This interface is critical for generating the Sun's large-scale magnetic fields.

1. **Cyclic Behavior:**

   The Sun's magnetic field follows an approximately 11-year cycle. During this cycle, the number of sunspots rises and falls, and the Sun's magnetic poles reverse. This cyclic behavior is a hallmark of the Sun's well-organized dynamo.

2. **Sunspots and Solar Activity:**

   Sunspots, flares, and coronal mass ejections are surface manifestations of the magnetic field generated by the dynamo. These phenomena have significant effects on the solar system, influencing space weather and Earth's magnetosphere.

How Dynamos Differ Between Stars

Magnetic dynamos are not universal; they vary significantly depending on a star's mass, size, composition, and stage of evolution.

1. Main-Sequence Stars

- **Small Stars (Red Dwarfs):**
    - Red dwarfs have fully convective interiors, meaning convection occurs throughout the entire star.
    - This structure allows for a highly efficient dynamo, resulting in strong, long-lasting magnetic fields.
    - Many red dwarfs exhibit intense starspots and flares due to this enhanced magnetic activity.

- **Sun-Like Stars:**
    - Stars like the Sun have a radiative core and a convective envelope, with the dynamo operating at the tachocline.
    - Their magnetic fields are relatively moderate and cyclic, as seen in the Sun's 11-year solar cycle.

2. Evolved Stars (Red Giants)

- **Structure of Red Giants:**
    - As stars like the Sun exhaust their core hydrogen and expand into red giants, their interiors change dramatically.
    - Red giants have a dense, inert core surrounded by a vast convective envelope.
    - The dynamo effect in red giants is less understood, but their deep convective layers are thought to sustain magnetic fields, even as their rotation slows.

- **Weak Magnetic Fields:**
    - In many red giants, the magnetic dynamo weakens due to slower rotation and the decoupling of the core and envelope.
    - However, XX Trianguli defies this trend by exhibiting strong and complex magnetic activity.

3. **Fast Rotators and Young Stars**

- **Enhanced Dynamos:**
    - Young and rapidly rotating stars generate powerful dynamos due to their high rotation rates and active convective zones.
    - These stars often exhibit chaotic, non-cyclic magnetic activity, with multiple poles and regions of intense magnetic flux.

- **Starspot Coverage:**
    - Fast rotators can have starspots covering a significant fraction of their surface, making them appear dimmer overall.

XX Trianguli: A Dynamo Outlier

XX Trianguli's magnetic dynamo operates in ways that challenge current models of stellar magnetism. Despite its slow rotation (approximately 400 days), it sustains a robust dynamo capable of producing strong magnetic fields and massive, persistent starspots.

Possible explanations include:

1. **Core-Envelope Interaction:**
   - The interaction between the contracting core and the expansive convective envelope may drive a dynamo effect unique to red giants.

2. **Residual Angular Momentum:**
   - XX Trianguli may have retained angular momentum from its earlier life stages, sustaining its dynamo longer than expected.

3. **Deep Convective Layers:**
   - The star's vast convective envelope provides an extensive region for magnetic field generation, compensating for its slow rotation.

The Implications of XX Trianguli's Non-Periodic Magnetic Activity

While the Sun's magnetic dynamo is cyclic and predictable, XX Trianguli exhibits **non-periodic magnetic activity**. This irregularity raises important questions about the mechanisms driving its magnetic field and the broader implications for stellar evolution.

1. Chaotic vs. Cyclic Magnetic Fields

XX Trianguli's magnetic behavior is highly chaotic, with no discernible cycle or periodicity in its activity. Starspots appear and evolve in unpredictable patterns, defying the regularity seen in the Sun's 11-year cycle.

Factors Contributing to Chaotic Activity:

- **Complex Magnetic Geometry:**
  - The Sun's magnetic field is predominantly dipolar, with a north and south pole. In contrast, XX Trianguli's field is multipolar, with numerous magnetic poles and regions of activity.

- **Irregular Rotation:**
    - Differential rotation in red giants is less well-understood than in main-sequence stars. Variations in rotational velocity across XX Trianguli's surface may contribute to its chaotic magnetic behavior.

- **Turbulent Convection:**
    - The star's convective envelope is vast and highly turbulent, creating a dynamic environment that destabilizes its magnetic field.

2. Long-Term Persistence of Starspots

One of the most intriguing aspects of XX Trianguli's magnetic activity is the long lifespan of its starspots. Some of these spots persist for years, far exceeding the lifespan of sunspots on the Sun.

Implications:

- **Stable Magnetic Anchoring:**
    - The magnetic fields associated with XX Trianguli's starspots may be anchored deep

within its convective envelope, preventing them from dissipating quickly.

- **Insights into Stellar Magnetism:**
    - Studying these persistent spots provides valuable information about the stability of magnetic fields in red giants.

3. Impact on Stellar Evolution

The magnetic activity of XX Trianguli has broader implications for its evolution:

1. **Mass Loss and Stellar Winds:**
    - Magnetic fields influence the star's mass-loss rate by shaping stellar winds.
    - Understanding XX Trianguli's magnetic field helps astronomers predict how it will lose mass and transition into its final stages as a white dwarf.

2. **Angular Momentum Loss:**
    - Magnetic activity can slow the star's rotation by transferring angular momentum to its stellar wind.

- The chaotic nature of XX Trianguli's activity suggests a less efficient mechanism for angular momentum loss compared to stars with cyclic fields.

4. Broader Implications for Stellar Populations

XX Trianguli's behavior challenges conventional models of red giants and may represent a subset of stars with unusual magnetic activity.

Insights for Other Stars:

- **Atypical Red Giants:**
  - Identifying other red giants with similar behavior could reveal whether XX Trianguli is an outlier or part of a larger class of stars.

- **Revised Dynamo Models:**
  - Current dynamo theories may need to be updated to account for the chaotic, non-periodic activity observed in stars like XX Trianguli.

5. Implications for Exoplanetary Systems

The magnetic activity of stars has significant effects on their surrounding planetary systems. For XX Trianguli, its chaotic magnetic behavior could influence:

1. **Habitability:**
   - Intense magnetic activity produces stellar flares and radiation that could strip atmospheres from nearby planets, reducing their habitability.

2. **Planetary Orbits:**
   - Magnetic fields and associated stellar winds can cause drag on planetary orbits, altering their positions over time.

Technological Advances and Future Research

The study of magnetic dynamos and their implications for stellar activity has been revolutionized by advancements in observational techniques.

1. Space-Based Telescopes

- Missions like **Kepler**, **TESS**, and the upcoming **PLATO** mission provide high-precision data on stellar brightness variations, enabling detailed studies of magnetic activity.

2. High-Resolution Spectrographs

- Instruments like **HARPS** and **ESPRESSO** offer unprecedented spectral resolution, allowing for precise measurements of magnetic fields and rotational velocities.

3. Computational Simulations

- Advances in computational modeling enable scientists to simulate the complex interactions between convection, rotation, and magnetic fields in stars like XX Trianguli.

The magnetic dynamo is the invisible engine driving stellar activity, from the familiar cycles of the Sun to the chaotic, unpredictable behavior of XX Trianguli. By studying this red giant, astronomers are uncovering new insights into the diversity of magnetic dynamos and their role in shaping the lives of stars.

XX Trianguli's non-periodic magnetic activity challenges established models, offering a rare glimpse into the complexities of red giant magnetism. As observational techniques and theoretical models continue to improve, the study of magnetic dynamos will remain at the forefront of astrophysics, illuminating the forces that govern the cosmos.

CHAPTER 8
The Impact of Stellar Activity on Surrounding Systems

"The life of a star is never confined to itself; its influence extends far beyond, shaping the environments of surrounding planets, moons, and the space in between."

Stellar activity is one of the most critical forces affecting a star's immediate environment. From the magnetic storms that generate solar flares to the colossal starspots on red giants like XX Trianguli, the effects of a star's magnetic behavior ripple outward, influencing surrounding systems in profound ways. The interplay between a star's magnetic activity and its environment has implications for the development and sustainability of planetary systems, the atmospheres of nearby planets, and even the potential for life.

In this chapter, we examine how magnetic activity influences a star's surroundings, with a particular focus on XX Trianguli. We explore the potential effects of its colossal starspots and chaotic magnetic fields on nearby planets, and we discuss the broader implications of stellar activity on planetary habitability and system dynamics.

How Magnetic Activity Influences the Star's Environment

The magnetic activity of a star generates a wide range of phenomena that extend well beyond its surface. These include stellar winds, magnetic fields, and high-energy radiation, all of which shape the physical and chemical conditions of the surrounding space.

1. Stellar Winds: Carriers of Magnetic Influence

Stellar winds are streams of charged particles ejected from a star's outer layers, often accelerated by its magnetic fields. These winds extend far into space, forming a **heliosphere**—a protective bubble around the star and its planetary system. However, the strength and behavior of stellar winds can vary dramatically depending on the star's magnetic activity.

Characteristics of Stellar Winds in Active Stars:

- **Increased Wind Intensity:**

    Stars with strong magnetic activity, like XX Trianguli, produce more intense stellar winds. These winds carry magnetic field lines outward, creating a highly dynamic stellar environment.

- **Magnetic Fields in Stellar Winds:**

    Magnetic fields embedded in stellar winds can interact with planetary magnetospheres, compressing or distorting them.

2. High-Energy Radiation

Stars with significant magnetic activity emit high-energy radiation in the form of X-rays and ultraviolet (UV) light. This radiation originates from the star's corona—a hot, magnetically active region extending above its surface.

Key Impacts of High-Energy Radiation:

- **Ionization of Planetary Atmospheres:**

    UV and X-ray radiation can ionize the upper layers of planetary atmospheres, leading to atmospheric escape over time.

- **Enhanced Space Weather:**

    High-energy radiation contributes to space weather events that can strip atmospheric particles from nearby planets.

In XX Trianguli, the chaotic and persistent magnetic activity likely produces heightened radiation levels compared to less active red giants, further influencing its surrounding system.

3. Magnetic Storms and Flares

Magnetic storms, driven by interactions between magnetic field lines on a star's surface, often result in stellar flares and coronal mass ejections (CMEs). These explosive events release massive amounts of energy and particles into space, disrupting the star's immediate environment.

Characteristics of Magnetic Storms in XX Trianguli:

- **Chaotic Behavior:**

    Unlike the Sun's predictable solar cycles, XX Trianguli's magnetic activity is non-periodic, making its flares and CMEs less predictable and potentially more intense.

- **Large-Scale Starspots as Triggers:**

    The massive starspots on XX Trianguli may act as focal points for magnetic reconnection, fueling powerful storms.

4. Astrophysical Jets and Outflows

For some highly active stars, magnetic fields can channel plasma into narrow jets that extend outward from the poles. While not directly observed in XX Trianguli, such outflows are common in young or magnetically active stars and may play a role in shaping the surrounding space.

Cumulative Effects on the Stellar Environment

The combined effects of stellar winds, radiation, flares, and jets create a highly dynamic and sometimes hostile environment. For stars like XX Trianguli, these forces are amplified by its magnetic complexity, influencing the structure of its heliosphere and the conditions in its planetary system.

The Potential Effects on Nearby Planets

The presence of nearby planets around XX Trianguli—or any magnetically active star—is an area of intense interest in astrophysics. Planets orbiting active stars are subject to a variety of influences, some of which may be beneficial, while others pose significant challenges to habitability.

1. Atmospheric Stripping

One of the most significant impacts of stellar activity on planets is the potential for atmospheric loss.

Mechanisms of Atmospheric Stripping:

- **Stellar Winds:**

    High-speed stellar winds can erode a planet's atmosphere, especially if the planet lacks a strong magnetic field.

- **Ionization by Radiation:**

    UV and X-ray radiation ionize atmospheric particles, making them more susceptible to being swept away by stellar winds.

Implications for Planets Around XX Trianguli:

The intense radiation and winds generated by XX Trianguli's chaotic magnetic activity would likely pose a significant challenge to the retention of planetary atmospheres, particularly for planets in close orbits.

2. Magnetosphere Compression and Disruption

A planet's magnetosphere acts as a shield against stellar winds and radiation. However, strong magnetic fields embedded in stellar winds can compress or distort this shield, exposing the planet's surface to harmful particles and radiation.

Effects on Planets with Weak Magnetic Fields:

- Increased surface radiation levels.
- Accelerated atmospheric loss.

Effects on Planets with Strong Magnetic Fields:

- Temporary auroras or magnetic storms.
- Enhanced protection, but with possible long-term weakening of the magnetosphere.

3. Climate Variability

Stellar activity can influence the climate of orbiting planets through changes in radiation output and space weather.

Examples:

- **Increased Flares:**

    Frequent flares can cause abrupt spikes in radiation, heating the upper atmosphere and potentially disrupting weather patterns.

- **Starspot Influence on Luminosity:**

    Large starspots can reduce a star's overall luminosity, leading to temporary cooling of planetary climates.

For XX Trianguli, its massive and persistent starspots may create irregular fluctuations in luminosity, potentially affecting the climates of any orbiting planets.

4. Tidal Forces and Orbital Instabilities

In some cases, magnetic activity and mass loss from a star can alter the gravitational forces within a planetary system, destabilizing planetary orbits over time.

For Planets Around XX Trianguli:

- Planets in close orbits may experience tidal interactions that lead to orbital decay.
- Mass loss through stellar winds could alter the star's gravitational pull, shifting the orbits of outer planets.

5. Impacts on Habitability

The potential habitability of planets depends on a delicate balance between receiving enough energy to sustain liquid water and avoiding conditions that strip away protective atmospheres or expose surfaces to harmful radiation.

Challenges Around XX Trianguli:

- The high-energy radiation and stellar winds from its chaotic magnetic activity would likely render close-in planets uninhabitable.

- Outer planets might escape direct exposure but could still be influenced by changes in the star's luminosity.

Possibilities for Resilience:

- Planets with strong magnetic fields and thick atmospheres might withstand the effects of stellar activity, preserving conditions suitable for life.

Case Study: Comparing XX Trianguli to the Sun

The Sun's relatively stable magnetic cycle provides a useful benchmark for understanding the unique challenges posed by XX Trianguli.

1. The Sun's Effects on Earth

- **Magnetic Protection:**

  Earth's magnetic field shields it from the worst effects of solar winds and flares.

- **Predictable Cycles:**

  The Sun's 11-year cycle allows for relatively consistent energy output, supporting stable climates.

2. Hypothetical Effects of XX Trianguli

If Earth orbited XX Trianguli, the following challenges would arise:

- **Increased Radiation Exposure:**

  Earth's magnetosphere might be insufficient to block the heightened radiation levels, leading to increased surface radiation.

- **Climate Instability:**

  Irregular starspot activity could cause extreme fluctuations in global temperatures.

Breakthroughs in Understanding Stellar Effects on Planets

The study of stars like XX Trianguli has significantly advanced our understanding of the relationship between stellar activity and planetary systems.

1. Exoplanet Discoveries

- Observations of exoplanets orbiting active stars have revealed the diverse ways stellar activity can shape planetary environments.

2. Space Missions

- **Kepler** and **TESS** have provided detailed data on starspot-induced brightness variations, offering insights into how active stars influence their planets.

3. Computational Modeling

- Simulations of star-planet interactions have improved our ability to predict the effects of magnetic fields and radiation on planetary atmospheres.

The Future of Research

As technology advances, our ability to study the effects of stellar activity on surrounding systems will continue to improve.

1. Next-Generation Telescopes

- Upcoming observatories like the **James Webb Space Telescope (JWST)** and the **Extremely Large Telescope (ELT)** will provide unprecedented insights into the atmospheres of exoplanets orbiting active stars.

2. Expanded Stellar Surveys

- Long-term monitoring of stars like XX Trianguli will help identify patterns in their magnetic activity and its influence on planetary systems.

The magnetic activity of stars like XX Trianguli extends far beyond their surfaces, shaping the environments of nearby planets and influencing their potential for habitability. From intense stellar winds to the dramatic effects of radiation and flares, the forces generated by stellar magnetic fields play a pivotal role in determining the fate of planetary systems.

For XX Trianguli, its chaotic magnetic activity and colossal starspots offer a unique opportunity to study these interactions in detail. As we continue to explore the cosmos, the lessons learned from stars like XX Trianguli will deepen our understanding of the intricate connections between stars and the planets that orbit them.

CHAPTER 9
Implications for Stellar Evolution

"The story of a star is written in cycles—of creation, transformation, and ultimate renewal. By studying stars like XX Trianguli, we glimpse the future of our Sun and the forces that shape the cosmos."

The study of stars provides one of the most fundamental keys to understanding the universe. Stars are the engines of cosmic evolution, creating and distributing the elements necessary for planets and life. As stars evolve, their physical and magnetic properties transform, shaping their environments and leaving an enduring legacy. Among aging stars, red giants like XX Trianguli offer profound insights into the final stages of stellar life cycles.

XX Trianguli stands out as an exceptional case study. Its colossal starspots, chaotic magnetic activity, and persistent variability make it a laboratory for understanding stellar aging. By examining the behavior and properties of XX Trianguli, scientists gain new perspectives on the processes that govern stellar evolution, from main-sequence stars to white dwarfs. This chapter explores what XX Trianguli teaches us about aging stars and how this knowledge deepens our understanding of stellar life cycles.

What XX Trianguli Teaches Us About Aging Stars

XX Trianguli is a red giant star, a phase that represents one of the most dramatic transformations in a star's life. Red giants are stars that have exhausted the hydrogen fuel in their cores, leading to profound structural and magnetic changes. By studying XX Trianguli, scientists are uncovering unique features of this evolutionary stage.

1. Changes in Stellar Structure

One of the defining characteristics of aging stars like XX Trianguli is the structural transformation they undergo as they transition from the main sequence to the red giant phase.

From Core Fusion to Shell Fusion:

- During the main sequence, stars sustain themselves by fusing hydrogen into helium in their cores.
- In red giants, the hydrogen in the core is depleted, causing the core to contract and heat up. Surrounding hydrogen in a shell begins to fuse, driving the star's dramatic expansion.

Structural Characteristics of XX Trianguli:

- **A Contracting Core:**

  The core of XX Trianguli is now composed mostly of helium, with fusion occurring in a surrounding hydrogen shell. This contraction contributes to the star's intense magnetic activity.

- **Expanding Outer Layers:**

  XX Trianguli has expanded to approximately 150 times the radius of the Sun, making its outer layers more diffuse and turbulent.

Implications for Stellar Evolution:

- These structural changes alter the star's internal dynamics, including convection patterns and rotation rates. Understanding how these shifts occur helps scientists model the evolution of stars at similar stages.

2. Magnetic Activity and Starspot Formation in Aging Stars

Red giants are traditionally thought to exhibit diminished magnetic activity due to their slower rotation. However, XX

Trianguli defies this expectation with its strong, chaotic magnetic fields and enormous starspots.

What Makes XX Trianguli Unique?

- **Persistent Magnetic Dynamo:**

    Despite its slow rotation, XX Trianguli sustains a robust magnetic dynamo, likely driven by its deep convective envelope and core-envelope interactions.

- **Colossal Starspots:**

    These spots, covering up to 30% of the star's visible surface, indicate that magnetic fields remain active and influential even in late stellar life.

Lessons for Stellar Magnetism:

- Magnetic activity in red giants is more diverse and persistent than previously believed.

- The study of XX Trianguli suggests that stars with unique internal dynamics may sustain magnetic fields longer than expected, challenging traditional models.

3. Mass Loss and Stellar Winds

Mass loss is a defining feature of the red giant phase. As stars expand, their outer layers become gravitationally unbound, creating powerful stellar winds that eject material into space.

Observations of XX Trianguli:

- **Enhanced Mass Loss:**

    XX Trianguli's magnetic activity likely contributes to its strong stellar winds, accelerating the ejection of mass.

- **Polarized Mass Loss:**

    Magnetic fields may channel mass loss into specific directions, creating asymmetries in the material surrounding the star.

Implications for Stellar Evolution:

- The material shed by red giants enriches the interstellar medium with heavy elements, fueling the formation of new stars and planets.

- Studying mass loss in stars like XX Trianguli helps refine models of how stars distribute elements back into the galaxy.

4. Variability and Instability

XX Trianguli exhibits irregular brightness variations caused by its massive starspots and pulsations in its outer layers. These fluctuations provide valuable information about the dynamic processes occurring within the star.

Key Observations:

- **Pulsations:**
  As the star expands and contracts, its luminosity varies, offering clues about its internal structure.

- **Starspot-Induced Variability:**
  The presence of large, persistent starspots adds another layer of complexity to the star's brightness variations.

Insights into Aging Stars:

- Variability in red giants is an indicator of their ongoing structural changes and the influence of magnetic activity.

- By studying these variations, astronomers can better understand the processes that drive the transition to the next stage of stellar evolution.

How This Research Contributes to Our Understanding of Stellar Life Cycles

The study of XX Trianguli not only reveals the unique characteristics of this star but also provides broader insights into the life cycles of stars, including their formation, evolution, and eventual demise.

1. The Transition from Main Sequence to Red Giant

The red giant phase marks a critical transition in a star's life. By observing XX Trianguli, scientists can trace the sequence of events that occur as stars exhaust their core hydrogen and begin burning hydrogen in a shell.

Key Contributions:

- **Understanding Core Dynamics:**

    The contraction of the core and the onset of shell fusion are critical steps in the red giant phase. Studying these processes in XX Trianguli provides data to refine theoretical models.

- **Magnetic Field Evolution:**

    The persistence of magnetic activity in XX Trianguli challenges traditional assumptions about the decline of magnetic fields in aging stars.

2. Mass Loss and Galactic Recycling

The material expelled by red giants like XX Trianguli plays a vital role in the galactic ecosystem. This ejected material contains heavy elements forged during the star's lifetime, enriching the interstellar medium and contributing to the formation of new stars and planets.

Contributions of XX Trianguli:

- **Refining Models of Mass Loss:**

    Observations of XX Trianguli's stellar winds provide valuable data on how mass is ejected and distributed in space.

- **Tracing Elemental Enrichment:**

    By studying the chemical composition of ejected material, scientists can better understand how red giants contribute to the galactic abundance of heavy elements.

3. The End of Stellar Evolution: Preparing for the White Dwarf Phase

As red giants shed their outer layers, they transition into the final stages of stellar evolution. For stars like XX Trianguli, this process culminates in the formation of a planetary nebula and a white dwarf core.

Lessons from XX Trianguli:

- **Pre-White Dwarf Dynamics:**

    Observations of variability and mass loss in XX Trianguli provide insights into the conditions that precede the formation of a white dwarf.

- **Magnetic Fields in White Dwarfs:**

    The persistence of magnetic activity in red giants raises questions about the role of magnetic fields in white dwarf formation and behavior.

4. Implications for Planetary Systems

The evolution of stars like XX Trianguli has profound implications for the planets that orbit them. As the star expands and sheds its outer layers, any surrounding planets are exposed to intense radiation, stellar winds, and gravitational changes.

Potential Effects on Nearby Planets:

- **Orbital Instabilities:**

  The loss of mass can alter the gravitational pull of the star, potentially destabilizing planetary orbits.

- **Atmospheric Stripping:**

  Intense radiation and stellar winds may strip away planetary atmospheres, rendering them uninhabitable.

- **Survival of Outer Planets:**

  Planets at greater distances may survive the red giant phase, but their environments are likely to change dramatically.

5. Insights into Stellar Populations and Galactic Evolution

XX Trianguli is part of a broader population of red giants that collectively shape the evolution of galaxies. By studying this star, scientists gain insights into the behavior of aging stars in different environments.

Broader Contributions:

- **Characterizing Red Giant Populations:**

    Observations of XX Trianguli help refine our understanding of the diversity within red giant populations.

- **Modeling Galactic Evolution:**

    The contributions of red giants to galactic chemical enrichment and star formation are critical for understanding the long-term evolution of galaxies.

Technological Contributions to Stellar Evolution Research

The study of stars like XX Trianguli has been made possible by significant advancements in observational technology and computational modeling.

1. Space-Based Observatories

- Telescopes like **Kepler**, **TESS**, and **Gaia** provide high-precision data on stellar variability, rotation, and composition.

2. High-Resolution Spectroscopy

- Instruments like **HARPS** and **ESPRESSO** enable detailed studies of stellar spectra, revealing information about surface activity and chemical composition.

3. Computational Simulations

- Simulations of stellar evolution allow scientists to model the processes occurring within stars like XX Trianguli, bridging the gap between observation and theory.

The Future of Stellar Evolution Research

The study of XX Trianguli and similar stars represents a frontier in astrophysics. As technology advances, scientists will gain an even deeper understanding of the processes that govern stellar life cycles.

1. Next-Generation Telescopes

- The **James Webb Space Telescope (JWST)** and the **Extremely Large Telescope (ELT)** will provide unprecedented resolution for studying red giants and their environments.

2. Expanded Surveys

- Large-scale surveys like **PLATO** will monitor thousands of stars, providing new insights into the diversity of stellar evolution.

3. Interdisciplinary Approaches

- Combining data from astrophysics, chemistry, and planetary science will provide a more holistic understanding of how stars influence their surroundings.

XX Trianguli offers a unique window into the life cycle of stars, illustrating the dramatic changes that occur as stars age and evolve. From its magnetic activity and mass loss to its variability and eventual fate, this red giant provides critical insights into the processes that shape stars and their environments.

By studying XX Trianguli, scientists are not only unraveling the mysteries of a single star but also deepening their understanding of the universal principles that govern stellar evolution. These lessons extend far beyond XX Trianguli, offering clues about the future of our Sun, the dynamics of galactic evolution, and the origins of the elements that make up the universe.

CHAPTER 10

Exploring the Universe Through Starspots

"In the darkness of the cosmos, starspots are not imperfections; they are keys to unlocking the mysteries of stellar behavior and the universe's grand design."

Starspots, once thought to be mere anomalies on the surfaces of stars, have proven to be invaluable tools for exploring the intricacies of stellar physics and the broader universe. Their presence reveals the unseen forces at work beneath the surfaces of stars and offers a glimpse into the invisible mechanisms that govern their evolution. Through stars like XX Trianguli, astronomers have turned these surface blemishes into a wealth of knowledge about magnetic fields, stellar life cycles, and the environments surrounding stars.

In this final chapter, we examine the broader significance of studying starspots and their role in understanding stellar behavior. We also look to the future of astronomy, exploring how research inspired by stars like XX Trianguli will shape upcoming scientific endeavors and deepen humanity's connection to the cosmos.

The Broader Significance of Studying Starspots and Stellar Behavior

Starspots are more than just a visual phenomenon; they are critical to understanding the internal dynamics of stars and the forces that govern their life cycles. Their study connects seemingly isolated phenomena, such as magnetic activity, stellar evolution, and planetary systems, into a cohesive understanding of the universe.

1. Unraveling Stellar Magnetic Fields

The presence of starspots is a direct manifestation of a star's magnetic field. By studying these spots, scientists can uncover the secrets of stellar magnetism and its role in shaping the behavior of stars.

Insights from Starspot Studies:

- **Dynamo Mechanisms:**

    Starspots are a direct result of the magnetic dynamo operating within a star. Studying their size, distribution, and persistence reveals how different types of stars generate and sustain their magnetic fields.

- **Magnetic Field Evolution:**

  Observing stars at various stages of their life cycles, like XX Trianguli, provides insight into how magnetic fields evolve over time. Red giants, in particular, challenge existing theories, as their slow rotation is typically associated with weak magnetic activity—yet XX Trianguli exhibits strong and persistent starspots.

- **Multipolar Magnetic Fields:**

  The complexity of XX Trianguli's magnetic field, with multiple poles and chaotic activity, has prompted astronomers to refine models of magnetic field generation and their effects on stellar surfaces.

2. Probing the Interiors of Stars

While starspots are surface phenomena, they provide indirect information about the internal structures of stars.

- **Core-Envelope Interaction:**

  In stars like XX Trianguli, interactions between the contracting core and the expansive convective envelope play a key role in generating magnetic fields. Starspot patterns offer clues about these internal processes.

- **Convection Zones and Rotation:**

    The distribution and size of starspots reflect the dynamics of a star's convective envelope and rotation rate. For red giants, this is particularly intriguing, as their large, turbulent convection zones create complex magnetic behavior.

3. Stellar Evolution and Life Cycles

Starspots are windows into the life cycles of stars. By observing starspots on stars at various stages of evolution, astronomers can trace the transitions from main sequence to red giant and beyond.

- **Youthful Activity:**

    Younger stars, such as T Tauri stars, are often highly magnetically active, with surfaces dominated by large, persistent starspots. Studying these stars reveals how magnetic activity diminishes with age.

- **Aging Stars:**

    Red giants like XX Trianguli demonstrate that magnetic activity can persist into late stellar life, challenging the assumption that older stars are magnetically quiescent.

- **End-of-Life Processes:**

    Understanding starspots on aging stars contributes to models of how stars lose mass and transition into white dwarfs, neutron stars, or black holes.

4. Planetary Systems and Habitability

Starspots influence not just the stars they adorn but also the environments of orbiting planets. By studying starspots, scientists can infer the conditions faced by exoplanets and assess their potential habitability.

Effects on Planetary Systems:

- **Radiation and Stellar Winds:**

    Starspots are often associated with heightened stellar activity, such as flares and coronal mass ejections (CMEs). These events can strip atmospheres from nearby planets or alter their climates.

- **Light Variability:**

    Starspots cause variations in a star's brightness, which can affect the energy received by orbiting planets, especially those in the habitable zone.

- **Exoplanet Detection:**

    Starspots are sometimes a source of noise in detecting exoplanets via transit methods. However, they also provide valuable data on stellar rotation and activity, which can be factored into exoplanet studies.

5. Contributions to Galactic Evolution

On a broader scale, starspots contribute to our understanding of how stars interact with their environments and influence the evolution of galaxies.

- **Mass Loss and Enrichment:**

    In red giants, magnetic activity and stellar winds shaped by starspots contribute to mass loss, enriching the interstellar medium with heavy elements necessary for the formation of new stars and planets.

- **Magnetic Fields in Galaxies:**

    Understanding stellar magnetic fields helps scientists model how magnetic fields influence the structure and dynamics of entire galaxies.

The Future of Astronomy and What XX Trianguli Inspires for Upcoming Research

The study of XX Trianguli and its colossal starspots is not just about understanding a single star—it's a stepping stone toward new discoveries and scientific frontiers. This star inspires a future of astronomy filled with advanced technologies, multidisciplinary collaborations, and deeper explorations of the cosmos.

1. Advancing Observational Technologies

The continued study of starspots will benefit from next-generation telescopes and instruments that push the boundaries of observational precision.

Key Developments:

- **High-Resolution Spectroscopy:**

    Instruments like **ESPRESSO** and the upcoming spectrographs on the **Extremely Large Telescope (ELT)** will provide unprecedented detail in studying stellar spectra, allowing for more accurate mapping of starspots.

- **Space-Based Telescopes:**

    The **James Webb Space Telescope (JWST)**, along with future missions like **PLATO**, will expand our ability to study stellar activity in distant stars and gather long-term data on their variability.

- **Interferometry:**

    Improved interferometric techniques may enable direct imaging of starspots on nearby stars, providing a visual confirmation of their size and distribution.

2. Enhanced Computational Modeling

As observational data becomes more detailed, computational models of stellar activity will need to keep pace.

- **Magnetic Field Simulations:**

    Simulations of dynamo processes in stars like XX Trianguli will help refine our understanding of how magnetic fields are generated and sustained in different stellar environments.

- **Starspot Evolution Models:**

    Tracking the formation, growth, and dissipation of starspots will offer insights into the underlying mechanisms driving magnetic activity.

3. Expanding Exoplanet Research

Starspot studies will play a critical role in the search for habitable exoplanets and the assessment of their environments.

- **Differentiating Starspots from Exoplanets:** Advanced algorithms are being developed to distinguish between starspot-induced brightness variations and the transits of exoplanets.

- **Planetary Habitability:**

    Research into the impact of stellar activity on planetary atmospheres and climates will inform models of habitability around active stars.

4. Exploring Red Giants and Beyond

XX Trianguli is a prototype for studying magnetic activity in red giants, but it also raises questions that extend to other stages of stellar evolution.

- **Late-Stage Stars:**

    How do magnetic fields behave as stars transition from red giants to planetary nebulae and white dwarfs?

- **Stellar Archetypes:**

    Are stars like XX Trianguli rare exceptions, or do they represent a broader class of magnetically active red giants?

5. Connecting Disciplines: A Multidisciplinary Approach

The study of starspots bridges multiple fields, including astrophysics, planetary science, and cosmology. Collaborative efforts will be essential for addressing the complex questions raised by stars like XX Trianguli.

- **Astrophysics:**

    Exploring the mechanisms of starspot formation and their connection to stellar dynamics.

- **Planetary Science:**

    Understanding how stellar activity influences the formation and evolution of planetary systems.

- **Galactic Evolution:**

    Investigating how stellar mass loss contributes to the chemical enrichment of galaxies.

6. Inspiring a New Generation of Astronomers

Stars like XX Trianguli not only expand scientific knowledge but also inspire curiosity and wonder. Their study serves as a gateway for future generations of astronomers, encouraging exploration and innovation.

7. The Philosophical Implications

Beyond its scientific significance, the study of starspots invites reflection on humanity's place in the universe. By understanding the forces that shape stars and planets, we gain a deeper appreciation for the interconnectedness of all things in the cosmos.

XX Trianguli, with its colossal starspots and chaotic magnetic activity, is more than an astrophysical curiosity—it is a beacon for the future of astronomy. Its study has advanced our understanding of stellar magnetic fields, life cycles, and their impact on surrounding environments. As we look to the future, the lessons learned from XX Trianguli will inspire new technologies, interdisciplinary collaborations, and groundbreaking discoveries.

Starspots, once viewed as surface imperfections, have become essential tools for unlocking the mysteries of the universe. They remind us that even the smallest features on the largest celestial bodies hold the power to illuminate the grandest questions of existence.

The universe beckons us to explore, and through stars like XX Trianguli, we move closer to understanding the forces that shape the cosmos and our place within it.

CONCLUSION
Handbook of Starspot Giants

"In the stars above us, we find not just light, but stories—stories of creation, transformation, and infinite possibilities. Starspot giants are among the greatest narrators of these cosmic tales."

The cosmos is a vast, dynamic tapestry woven from the lives of stars. Within this grand framework, the phenomenon of starspots offers a fascinating and unparalleled glimpse into the inner workings of stars. These magnetic blemishes, colossal in size and profound in significance, are far more than mere surface features. They are windows into the magnetic and dynamic processes that define a star's life cycle and impact its surrounding environment.

As we reach the conclusion of *Handbook of Starspot Giants*, it becomes clear that starspots are not just curiosities—they are powerful tools for understanding stars, their evolution, and their role in the broader universe. Through the study of stars like XX Trianguli, we have uncovered patterns and behaviors that challenge long-held assumptions and inspire new avenues of research. This book has explored their significance in exquisite detail, and here, we bring together the threads of these insights into a cohesive understanding.

Starspot Giants: Cosmic Laboratories

Stars like XX Trianguli are more than extraordinary examples of starspot activity—they are cosmic laboratories where the principles of astrophysics play out on an epic scale. The discovery and study of colossal starspots have opened new doors in understanding the forces that shape stellar magnetism and surface phenomena.

1. A Deep Dive Into Stellar Magnetism

- **Magnetic Dynamos:**

    At the heart of every starspot lies a magnetic field born from the dynamo effect. The interaction of a star's rotation and convection generates these fields, and their presence manifests as starspots on the surface.

- **Complexity in Red Giants:**

    Stars like XX Trianguli demonstrate that even in the later stages of stellar evolution, magnetic fields can remain strong and dynamic. This challenges previous assumptions that red giants are magnetically quiescent, expanding our understanding of how magnetic dynamos operate in aging stars.

2. The Role of Starspots in Stellar Evolution

Starspots are more than just localized surface features—they are intimately tied to the life cycle of a star. From their formation in rapidly rotating young stars to their persistence in aging giants, starspots serve as indicators of a star's internal processes and external interactions.

- **Tracing Stellar Lifetimes:**

  Starspots allow us to study the internal mechanics of stars at various stages, offering clues about their past, present, and future.

- **Insights Into Late-Stage Evolution:**

  In stars like XX Trianguli, persistent and chaotic starspots provide vital information about the transition from main sequence to red giant and ultimately to white dwarf.

The Impact of Stellar Activity Beyond the Star

The study of starspots also underscores how interconnected the universe truly is. A star's magnetic activity, as evidenced by its starspots, has far-reaching consequences, affecting not only the star itself but also its surrounding environment.

1. Planetary Systems at Risk

- **Radiation and Stellar Winds:**

    Magnetic activity often correlates with increased stellar flares and winds, which can strip planetary atmospheres or alter their climates. Planets orbiting active stars like XX Trianguli would face extreme conditions that could challenge their habitability.

- **Variable Luminosity:**

    Starspots cause brightness fluctuations, which can impact the energy received by orbiting planets, influencing their weather and potentially their ability to sustain life.

2. Galactic Contribution

Stars like XX Trianguli play a vital role in enriching the interstellar medium. Their mass loss, driven by magnetic fields and stellar winds, releases heavy elements into space, fueling the formation of new stars and planets. The study of such stars provides critical insights into the chemical evolution of galaxies.

The Technological Revolution in Starspot Studies

Over the past few decades, advancements in technology have transformed the study of starspots from a niche field into a cornerstone of astrophysics. Space-based telescopes, high-resolution spectroscopy, and advanced computational models have allowed scientists to probe the surfaces of stars in ways that were once thought impossible.

1. Pioneering Techniques

- **Doppler Imaging:**

    By detecting Doppler shifts in light emitted by rotating stars, astronomers have created detailed maps of starspot distributions, unlocking new insights into their size and behavior.

- **Spectropolarimetry:**

    This technique has been critical for studying the magnetic fields associated with starspots, revealing the intricate interplay between magnetic activity and stellar dynamics.

2. Data From Missions

Space telescopes like **Kepler**, **TESS**, and the upcoming **PLATO** mission have been instrumental in monitoring stellar brightness variations, providing invaluable data on starspots and their evolution. With the continued development of instruments like the **James Webb Space Telescope (JWST)** and the **Extremely Large Telescope (ELT)**, the future of starspot research looks brighter than ever.

Challenging Long-Held Assumptions

Perhaps one of the most significant contributions of starspot research is its ability to challenge and refine existing models of stellar behavior. The findings from stars like XX Trianguli have forced astronomers to reconsider their understanding of stellar magnetism and life cycles.

1. Magnetic Fields in Aging Stars

It was long believed that red giants, due to their slower rotation rates, would exhibit weak or negligible magnetic fields. The discovery of strong, persistent magnetic activity in XX Trianguli demonstrates that this is not always the case. Instead, red giants can sustain complex magnetic fields through interactions between their cores and convective envelopes.

2. Non-Cyclic Magnetic Behavior

Unlike the Sun's predictable 11-year solar cycle, the magnetic activity in stars like XX Trianguli is chaotic and non-periodic. This irregularity suggests that dynamo mechanisms in aging stars are more diverse than previously thought, opening new avenues for theoretical research.

The Future of Starspot Research

As our understanding of starspots deepens, so too does our appreciation for their broader implications. The study of starspots is no longer confined to individual stars—it is a gateway to understanding galaxies, planetary systems, and even the origins of life.

1. Expanding Observations

The continued observation of starspots across diverse types of stars will help refine models of stellar evolution. Future missions will monitor stars at every stage of their life cycle, providing a more complete picture of how magnetic activity evolves over time.

2. Multidisciplinary Collaboration

The study of starspots bridges multiple fields, from astrophysics and planetary science to space weather and cosmology. Collaborative efforts will ensure that insights gained from starspot research are applied across disciplines, enriching our understanding of the universe.

3. New Questions to Answer

- How common are magnetically active red giants like XX Trianguli?
- What role do magnetic fields play in the final stages of stellar evolution?
- How do starspots influence the long-term habitability of planets?

These questions, inspired by stars like XX Trianguli, will guide the next generation of astronomical research.

A Final Reflection on Starspot Giants

The study of starspots, and especially those on giants like XX Trianguli, reveals the profound beauty and complexity of the universe. These seemingly minor features hold the keys to understanding some of the most fundamental processes in nature, from the generation of magnetic fields to the recycling of stellar material into new stars and planets.

As we continue to explore the cosmos, stars like XX Trianguli remind us of the interconnectedness of all things.

The forces at play within a single star ripple outward, influencing entire galaxies and shaping the environments in which planets—and life—exist.

Looking to the Stars

Stars have always been humanity's companions in the night sky, inspiring curiosity and wonder. Through the study of starspots, we move beyond admiration to understanding, uncovering the forces that govern the universe and our place within it. *Handbook of Starspot Giants* celebrates this journey of discovery, offering a glimpse into the lives of stars and the lessons they teach us.

As we gaze upward, let us remember that the stars are not static—they are dynamic, evolving, and alive with magnetic forces that shape their stories. Through them, we find the answers to questions both cosmic and personal, and we are reminded of the infinite possibilities that await us in the universe.

Let *Handbook of Starspot Giants* serve as a guide, a resource, and an invitation to continue exploring the stars, where every spot, every flicker, and every light carries the promise of discovery.

www.ingramcontent.com/pod-product-compliance
Lightning Source LLC
Chambersburg PA
CBHW071057240526
45471CB00016B/1972